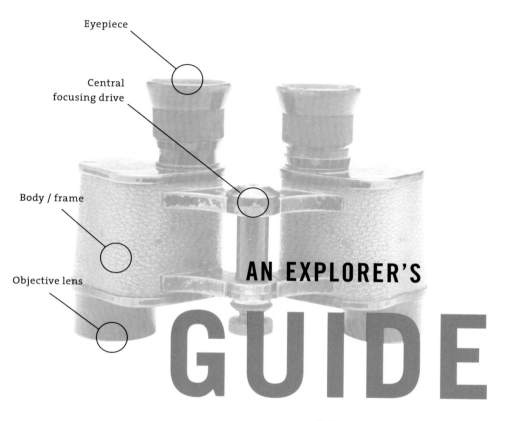

Eyepiece

Central focusing drive

Body / frame

Objective lens

AN EXPLORER'S GUIDE

TO
THE FIELD MUSEUM

LOGAN WARD

NATURE

1. Martian Meteorites
2. Chalmers Topaz
3. Lapis Lazuli
4. Fall of the Benld Meteorite
5. Mazon Creek Fossils
6. Dimetrodon
7. Charles Knight Mural
8. Parasaurolophus
9. Fossil Lake
10. McDonald's Fossil Prep Lab
11. Sue
12. Earwig Nest
13. Tropical Aerial Garden
14. Common Inky Cap Mushroom
15. Welwitschia
16. Amazon Victorian Water Lily
17. Cacao
18. California Condor
19. Local Woodlands Four Seasons Diorama
20. Great Horned Owl
21. Blue Bird of Paradise
22. Pink Fairy Armadillo
23. Black Right Whale Skeleton
24. Argali
25. Mexican Grizzly Bear

TABLE OF CONTENTS

inside front cover
MAP OF THE FIELD MUSEUM

26 Bushman
27 African Elephants
28 The Lions of Tsavo
29 Marsh Birds of the Upper Nile River
30 Elephant Dung Beetle

CULTURE

31 Living Together Shoes
32 Predynastic Egyptian Burial
33 Egyptian Mastaba
34 Egyptian Inner Coffin of Chenet-a-a
35 Egyptian Funeral Boat of Sen-Wosret III
36 Egyptian Cat Bronze
37 Bakongo Nkondi Figure
38 Benin Queen Mother's Memorial Head
39 Tuareg Saddlebag
40 Pawnee Earth Lodge
41 Hopi Great Horned Owl Kachina Dolls
42 Apache Jar-Shaped Coil Basket
43 Hopewell Mica Bird Claw
44 Tlingit Halibut Hook
45 Haida (Skidegate) Potlatch Copper
46 Point Hope Shaman Mask
47 Kwakiutl Transformation Mask
48 Bella Coola Grizzly Bear House Post
49 Aztec Diorite Boulder with Quetzalcoatl Figure
50 Chinese Jade Jar
51 Chinese Porcelain of Guanyin
52 Chinese Christian Madonna and Child
53 Japanese Inro
54 Tibetan Silver Deity, Chenrezi
55 Jaluit Atoll Diorama
56 Ruatepupuke II: A Maori Meeting House
57 Polynesian God Figure
58 Melanesian Rambaramp
59 Melanesian Pig Mask
60 Malvina Hoffman Zulu Woman and Padaung Woman

inside back cover
MAP OF THE WORLD

May 2, 1921, opening day at The Field Museum's present location, just south of downtown Chicago.

INTRODUCTION

THE FIELD MUSEUM'S MISSION – exploring the Earth and its people – becomes apparent as soon as you approach the Museum. Outside, a life-sized cast of a *Brachiosaurus* stands three stories tall. Just inside the building's bronze doors, two Haida Indian totem poles rise to the ceiling. Beyond them, a pair of African Elephants are locked in battle. And at the far end of Stanley Field Hall, the fossilized bones of Sue, the largest and most complete *Tyrannosaurus rex* ever found, are frozen mid-stride.

These are just a few of the exciting discoveries you can make as you explore The Field Museum. The Museum's cultural, archeological, botanical, zoological, and geological collections total more than 20 million objects, less than one percent of which is on display. As a way to introduce visitors to this vast collection, we have highlighted in this guidebook 60 objects representing the diversity of what you will find in The Field Museum's halls.

This guidebook is divided into two categories: nature and culture. Plants, animals, rocks, and fossils fill the first half of the book. Objects of human culture from Asia, the South Pacific, Egypt and other parts of Africa, and North and South America make up the other half. Each object is assigned a number that is keyed to two maps:

a front-cover overview of the Museum's floorplan and a back-cover map that shows the part of the world from which each of the objects originates. (Keep in mind that because of conservation needs, changes in the Museum's installations, or exhibition loans, certain items featured in this book may be temporarily off display.)

AN INTEGRATED APPROACH. What you will see as you explore the Museum, however, is only part of what The Field Museum is all about. Inseparably linked to the exhibits are research and education. From the rain forests of southern Chile, to Wyoming's fossil lakes, to the islands of the South Pacific, The Field Museum supports the studies of more than 70 scientists. Their work is helping to address such vital concerns as species extinction, human conflict, and diseases like cancer and AIDS.

As the scientists learn about the world, so does the public, thanks in large part to the efforts of The Field Museum's education and exhibits departments. Through libraries, lectures, field trips, courses, and other programs, the Museum touches the lives of close to a million-and-a-half people every year, many of whom are children.

HOW IT ALL BEGAN. This threefold approach – research, education, and exhibits – has driven our operation since the Museum was founded. After the success of the 1893 World's Columbian Exposition, a fair held in Chicago to celebrate the 400th anniversary of Christopher Columbus's voyage to the Americas, a group of Chicagoans decided to create a natural history museum using some of the objects from the exposition. That same year, with major funding from Marshall Field, founder of Chicago's most famous department store, they started what was then known as the Field Columbian Museum.

It occupied the Palace of Fine Arts, a building on the south side of Chicago that was left over from the fair. In 1905 the name was changed to The Field Museum of Natural

History, and in 1921, with more funding bequeathed by Mr. Field, the Museum completed construction on its current home, a white-marble-clad neoclassical structure designed by renowned architect Daniel H. Burnham.

Marshall was not the only Field to leave his mark on the Museum. From 1909 to 1962 his nephew Stanley Field served as the Museum's president, expanding the collections and personally navigating the Museum through decades of change. It was Stanley Field who commissioned two of the Museum's great bodies of original art – Malvina Hoffman's sculptures of people from around the world, and Charles Knight's murals of prehistoric life on Earth. Another Marshall Field, the founder's great, great grandson, is now an active member of the Board of Trustees.

A DYNAMIC INSTITUTION. Today the Museum is as committed as ever to facilitating the understanding of the great diversity of human culture and the natural environment, past and present. Always changing, it continues to focus on the two most important issues of our time: a more prudent stewardship of our environment and improved cultural understanding among people – themes you will see in every exhibit. We hope this guidebook will enhance your journey toward these goals and through the halls of The Field Museum.

JOHN W. McCARTER, JR.
President and CEO
The Field Museum

Field Museum scientist next to *Brachiosaurus* bones in plaster, Grand Junction, Colorado, 1900.

NATURE

Scientists in the **BOTANY DEPARTMENT** at The Field Museum study plants and fungi. The botany collection, totaling 2.6 million carefully preserved specimens, is the fifth largest herbarium in the Western Hemisphere. This collection covers all major plant groups and every continent on Earth. It is especially rich in flowering plants and ferns of the Americas, and mosses, liverworts, and fungi of the world.

The primary focus of The Field Museum's **GEOLOGY DEPARTMENT** is paleontology, with seven of its eight curators covering fossil animals and plants. Accordingly, the Museum's collections of fossil mammals, fossil fishes, and fossil invertebrates are among the largest in the world. Most of the department's paleontologists take an interdisciplinary approach to their research, studying both fossil and living organisms to better understand the big picture of evolution. One curator specializes in the study of meteorites, and our collection of meteorites ranks among the top three in the nation.

The **DEPARTMENT OF ZOOLOGY** is the largest of the Museum's four curatorial departments. It is organized into six divisions: Birds, Fishes, Insects (with Arachnids), Amphibians and Reptiles, Invertebrates, and Mammals. The Museum's collections total more than 17 million zoological specimens and rank among the world's largest and most significant.

Upper West

Upper East

While there are thousands of meteorites in museums and collections around the world, only 12 of them are believed to be from Mars. Five of these are part of The Field Museum's collection, two of which, Zagami and Lafayette, are on display.

Why do scientists think they are from the Red Planet? One piece of evidence is that the trapped gases in the meteorites match those in Mars's atmosphere, which the Viking spacecraft measured in 1976. Scientists theorize that two great asteroid collisions on Mars hurled clouds of rubble into space; the Lafayette rock was ejected after an impact 11 million years ago, and Zagami was the result of an asteroid crash 2.5 million years ago. Drawn by the gravitational pull of our planet, these two Mars fragments eventually fell to the earth. According to eyewitnesses, one landed in the town of Zagami, Nigeria, in 1962. No one knows exactly where the other hit. In 1931 O. C. Farrington, a Field Museum curator, found it in the geology collection of Purdue University in West Lafayette, Indiana.

Meteorites are our only rock samples from Mars, and they provide clues about the history of the planet. They have shown that there was at one time water on Mars, possibly as "recently" as 180 million years ago. In 1996 a group of scientists identified what they claim is fossilized nanobacteria on a Martian meteorite found in Antarctica, which might be considered evidence that life exists or has existed elsewhere in the solar system.

MARTIAN METEORITES 1

At 5,890 carats, this avocado-size gem, purchased in 1966 with a fund established by William and Joan Chalmers, is the world's largest cut blue topaz. Found in a stream in Minas Gerais, Brazil, it was later faceted. Topaz exists in various colors: from almost colorless to golden-orange, yellow, brownish-yellow, pink, red, and various shades of blue. The most valuable, called imperial topaz, is golden-orange. Topaz registers an 8 on Mohs's scale, which measures hardness from 1 (talc) to 10 (diamonds). Topaz is the birthstone for November.

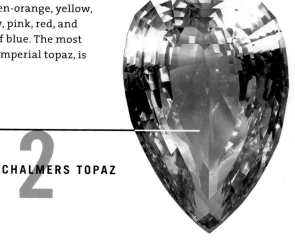

2 CHALMERS TOPAZ

Mystery surrounds this 312-pound (706,000-carat) block of lapis lazuli, one of the world's largest ever found. In the late 1800s, the deep-blue semiprecious boulder was discovered in an Inca Indian grave in Peru. But, curiously, the nearest known source for such a stone is more than 600 miles to the south, in Chile. And the quality and color of this 24" x 13" x 10" specimen is superior to Chilean lapis. Did the Incas somehow transport the huge stone to Peru? Is there a lost Inca lapis lazuli mine somewhere in the Peruvian Andes? No one knows for sure.

Most of the world's lapis today comes from mines in Afghanistan. Believing it to have curative powers, ancient people sometimes crushed the blue stone into a fine powder, mixed it with milk, and drank it. Others rubbed the dust into open wounds.

LAPIS LAZULI 3

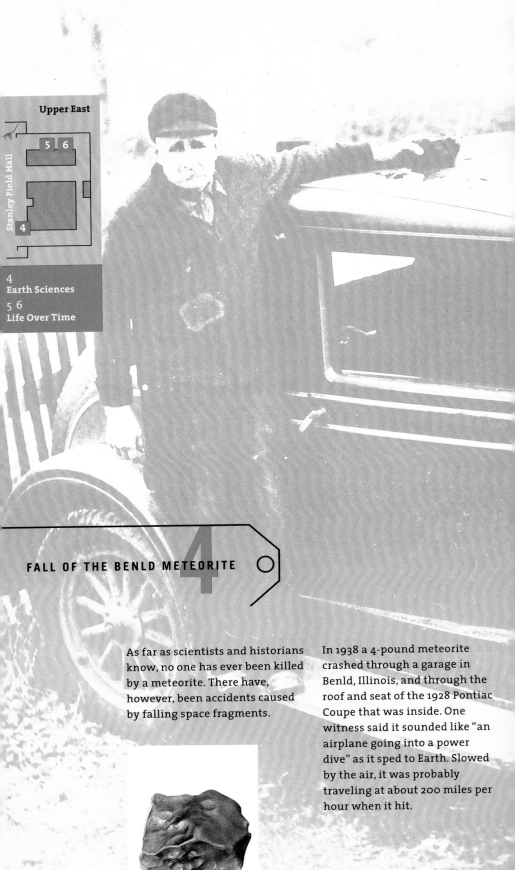

Upper East

4 Earth Sciences
5 6 Life Over Time

FALL OF THE BENLD METEORITE

As far as scientists and historians know, no one has ever been killed by a meteorite. There have, however, been accidents caused by falling space fragments.

In 1938 a 4-pound meteorite crashed through a garage in Benld, Illinois, and through the roof and seat of the 1928 Pontiac Coupe that was inside. One witness said it sounded like "an airplane going into a power dive" as it sped to Earth. Slowed by the air, it was probably traveling at about 200 miles per hour when it hit.

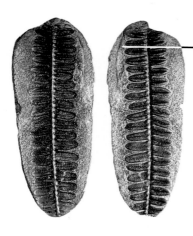

MAZON CREEK FOSSILS 5

Like the richest gold and diamond deposits, there exist throughout the world important concentrations of fossils. Among them are the Mazon Creek fossils of northeastern Illinois, which provide one of the most complete records of late Paleozoic life, roughly 300 million years ago. The Mazon Creek fossils are found in ironstone concretions, or hard masses, about the size of your fist that are embedded in soft gray shale. Over the years, as the water in Mazon Creek cut into these shale banks, it loosened the concretions and pitched them to the riverbed. Today, most lie in spoil piles left by coal mining companies. Crack one open and you might find a pristine, three-dimensional record of a fern leaf, worm, clam, fish, or even a mystery creature called Tully Monster. The Field Museum has thousands of Mazon Creek fossils in its collection.

DIMETRODON 6
Dimetrodon grandis

Believe it or not, this fin-backed creature was more closely related to humans than dinosaurs. It lived during the early Permian period, 270 million years ago, in what is today Texas. Part of the order Pelycosauria, it was among the world's first synapsids, animals characterized by the existence of a single hole in the skull behind each eye socket. This hole, which *Homo sapiens* also has, evolved as an attachment point for sophisticated jaw muscles. Until then, skulls were solid and jaws were little more than snapping devices. At the time, *Dimetrodon*, which grew up to 10 feet in length, was one of the largest and fiercest predators on Earth. Lining its powerful jaws were rows of serrated teeth for slicing through flesh, with large, daggerlike incisors in the front for stabbing (its name means "two-sized tooth"). Its large dorsal fin may have been another advantageous physical feature. The cold-blooded *Dimetrodon*, scientists believe, used the broad flap of skin as a solar panel to warm its blood more quickly and thus to give it a jump each morning over its more sluggish prey.

Charles Knight, influenced by his mentor, Henry Fairfield Osborn, head of New York's American Museum of Natural History, painted the *Apatosaurus* (formerly known as the *Brontosaurus*) with the thick, stubby head of the *Camarasaurus*. In the late 1970s scientists at the Carnegie Museum in Pittsburgh proved Osborn wrong by locating a real *Apatosaurus* head in their collection. The Field Museum was the first to receive a cast, and today it sits at the end of the *Apatosaurus*'s long neck in the Life Over Time exhibit. You will see the old skull standing beside the dinosaur – and Charles Knight's interpretation of what it might have looked like in his mural.

CHARLES KNIGHT MURAL
Apatosaurus, a Brontosaurus

If you attended school in the United States at any time since the 1930s, chances are you have seen one of Charles Knight's 28 murals of prehistoric life, which were commissioned by The Field Museum in 1927. These vivid, detailed depictions of dinosaurs, Ice-Age mammals, and early humans have been reproduced countless times in science textbooks. Yet while new scientific findings have rendered some of Knight's paintings scientifically obsolete, the huge canvases have retained their drama and beauty.

Upper East

7
Life Over Time, Elizabeth Morse Genius Dinosaur Hall

Although it was found in New Mexico in 1923, this *Parasaurolophus* really was not "discovered" until the 1950s. At that time, John Ostrom, a paleontologist visiting from Yale University, realized that the fossils sitting for decades in the Museum's third-floor storage area were from a previously unknown species. This is known as a "holotype," or type specimen, the first of its kind identified and the specimen to which other *Parasaurolophus cyrtocristatus* bones will always be compared.

This dinosaur is from the late Cretaceous period, 65 to 90 million years ago. Its parts were to remain accessible for study by paleontologists from around the world. Eventually, however, the Museum created a special display that carefully cradles the bones but also allows scientists and students to remove each one separately for study. There may even be vertebrae or other bones missing from the mount when you visit.

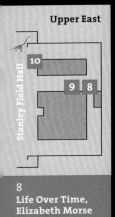

Upper East

8
Life Over Time,
Elizabeth Morse
Genius Dinosaur
Hall

9
Life Over Time

10
Stanley Field Hall

PARASAUROLOPHUS 8
Parasaurolophus cyrtocristatus

Scientists theorize that the curved head crest may have been used to trumpet warning or mating calls.

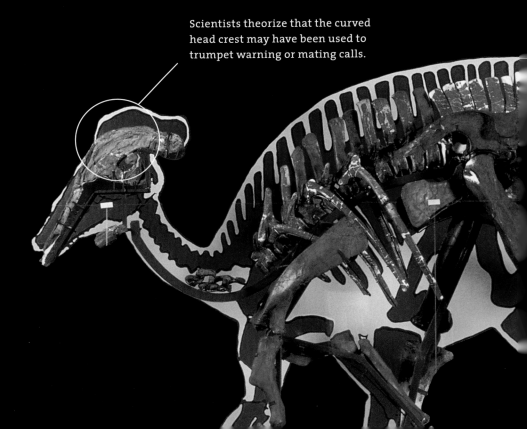

Scientists have not always understood fossils. Some people thought they had grown from seeds dropped by stars or had been carved by prehistoric artists. Others considered fossils the petrified bodies of demons pushed from the underworld. Today fossilized plants and animals help fill in the gaps of evolution by providing a record of life millions of years ago. From this well-preserved fish (*Priscacara serrata*), for instance, paleontologists can study bone structure, fins, and even stomach contents – just like biologists do with species living today – which allows them to categorize the fish and learn about its ecological history. It is one of thousands of fishes in the Museum's Green River Formation collection, the largest such collection in the world. The rocks of the formation come from sediment deposited by three Tertiary lakes (44 to 55 million years old) in what is now Wyoming, Colorado, and Utah. By cutting and peeling back broad, thin slabs of limestone from one of the extinct lakes – known as Fossil Lake – paleontologists have retrieved layer upon layer of fossil fishes, reptiles, birds, plants, and other organisms. One stone from the region was found to contain more than 200 different fish skeletons.

9 FOSSIL LAKE

10 MCDONALD'S FOSSIL PREP LAB

This state-of-the-art lab opened in 1998 to give visitors an insider's look at some of the scientific work that's done here at The Field Museum. Visitors can watch fossil preparators as they chip rock away from fossils, revealing traces of plants and animals that lived millions of years ago.

Preparators play an important role in making fossils available for scientific study and public display. For example, when Sue the *T. rex* first arrived at the Museum, most of her bones were encased in solid rock. For nearly two years a team of preparators worked full-time in this lab to clean and prepare Sue's bones.

Today preparators work on a variety of fossils in this lab. Their projects range from dinosaur bones that were collected in the early 1900s but were never prepared to the latest fossil discoveries made by Field Museum scientists.

Main East

11 Stanley Field Hall
12 Underground Adventure

Ground East

What makes this dinosaur so special? Simply put, Sue is the largest, most complete, best preserved, and most famous *Tyrannosaurus rex* ever found. And *T. rex* is the largest, most ferocious meat-eater ever to roam North America.

Sue the *T. rex* is named after Sue Hendrickson, the fossil hunter who discovered the dinosaur's massive bones in the Badlands of South Dakota. Although we often refer to Sue as a "she," we have no way of knowing if Sue – or any other *T. rex* – was male or female.

Since Sue first arrived at The Field Museum in 1997, her 67-million-year-old bones have received a lot of attention. A team of 10 fossil preparators spent over 25,000 hours cleaning and repairing each of Sue's more than 200 bones. Museum researchers examined, identified, measured, and described every detail of these bones. In addition, Sue's five-foot-long skull underwent a full industrial CT scan so that scientists could study the skull inside and out – without having to saw it in half.

All of this attention is yielding new information about how *T. rex* lived and died. CT scans show that sense of smell was extremely strong in this species. And medical X-rays indicate that Sue suffered – and survived – broken bones and other injuries. Who knows what other discoveries await future generations of paleontologists who study this remarkable *T. rex* specimen.

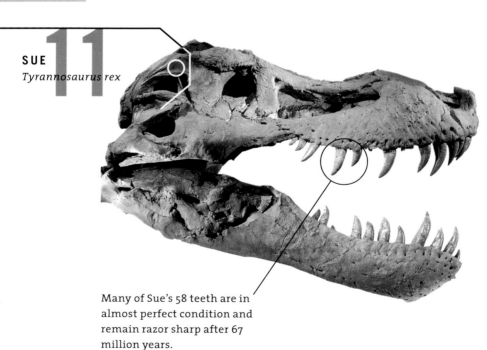

SUE 11
Tyrannosaurus rex

Many of Sue's 58 teeth are in almost perfect condition and remain razor sharp after 67 million years.

This mother earwig is very protective of her wriggling brood. Get too close and she will rear up and menacingly click her 18-inch pincers.

EARWIG NEST
Underground Adventure

The earwig is one of 10 larger-than-life animatronic creatures that can be seen in the Underground Adventure exhibition, where visitors are shrunk to 1/100th their normal size. Because so many soil organisms are far too small to see with the naked eye, experiencing them at such a large size helps us understand the abundance and diversity of life in the amazing world that's just beneath our feet.

More than 3,500 models – all patterned after real specimens in The Field Museum's collections – appear in this soil environment. To make the models, highly magnified photographs were taken of each specimen with an electron microscope. Then model-makers worked with the real specimens, the photographs, and the Museum's scientists to create accurate replicas for the exhibition.

This root system reaches from floor to ceiling in the Underground Adventure exhibition.

What do these tropical hanging plants have in common with false teeth? Like most of the other plant reproductions at The Field Museum, they were made between 1909 and 1969 by artisans at the Museum's model facility, whose first director was trained as a denture maker. His name was B. E. Dahlgren, and along with Charles Frederick Millspaugh, the Museum's first botany curator, he helped revolutionize the way museums display plants. Instead of filling glass cases with timber, dried leaves, seeds, and other nonperishable plant material, Dahlgren developed a way of molding wax, glass, and wire (and, after World War II, plastic) into lifelike stems, leaves, fruits, and flowers. Take a closer look at some of the features – the bromeliad's pink flowers, the green cactus leaves – and imagine the attention to detail that went into making each one.

13 TROPICAL AERIAL GARDEN

The cap of this mushroom, found commonly in Illinois, dissolves into a dark substance that so resembles ink, some artists actually dip their brushes into it and paint with it. Scientists believe the phenomenon is the mushroom's way of propagating itself: As the edge deteriorates into the inky substance, new spores (single-celled particles that can develop into a new mushroom) are cast into the air and spread by the wind. Many trees disperse seeds the same way. But mushrooms are not plants; they are part of a different kingdom of organisms called fungi. One important difference is that plants gather water, minerals, and other nutrients from the soil and make their own food using photosynthesis; fungi absorb all of their sustenance from other organisms around them. In fact, people have been known to get sick from eating edible mushrooms growing near yards recently treated with pesticides.

14 COMMON INKY CAP MUSHROOM
Coprinus micaceus

Charles Darwin called this bizarre-looking species "the duckbilled platypus of the plant world." Dr. Joseph Hooker, director of England's Royal Botanic Gardens from 1866 to 1885, said, "It is without question the most wonderful plant ever brought to [the United Kingdom] – and the very ugliest." Found only in the Namib Desert of southwest Africa, Welwitschia is definitely unique: It can live for hundreds of years (carbon-14 dating proved one specimen survived nearly 2,000 years); during that long lifetime it grows only two leaves, which split as they are ravaged by the harsh desert winds; and the tough, fibrous surface of its leaves reflects up to 50 percent of the sun's harmful rays, keeping the plant from overheating. The only species in its genus, *Welwitschia mirabilis* was discovered in 1859 by Austrian botanist Friedrich Welwitsch. Scientists today are still not sure how it obtains the water it needs in a desert that can go without rain for 10 years. Some say it lives off of dew.

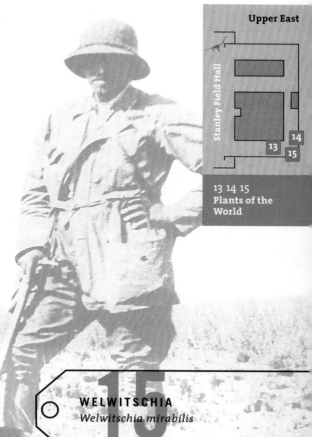

Upper East

13 14 15
Plants of the World

15

WELWITSCHIA
Welwitschia mirabilis

Professor H. Humbert of the Natural History Museum of Paris, while collecting specimens for The Field Museum, southwestern Africa, 1937.

Plants sometimes play a surprisingly active role in transferring their own pollen. Take, for instance, this giant water lily, the world's largest aquatic plant.

Among the people of the Amazon River region where the lily grows, the plant is known as the *forno de jacana*, or the jacana's frying pan. The jacana is a long-legged wading bird that forages among its leaves.

AMAZON VICTORIAN WATER LILY
Victoria amazonica

16

In the evening the lily's flowers open and lure beetles inside with small food particles. Then they close, trapping the insects until the next evening – plenty of time for one of the creatures inadvertently to dust its legs with pollen before leaving and carrying the pollen to another flower.

The Victorian lily's floating, skilletlike leaves are broad and sturdy enough to hold a small child.

Candy bars, hot cocoa, and boxes of Valentine sweets – all things chocolate – come from the seeds of this tropical rainforest plant. But you wouldn't guess that to slice open one of the cacao tree's drooping pods and dig out its seeds. First of all, they're encased in a sweet-and-sour pulp that, although it is considered a treat in the places cacao grows, has nothing in common with the rich, creamy taste of chocolate. And the acorn-size beans themselves are not sweet, either. Both the Mayan Indians, who worshipped cacao, and the Aztecs, who used the beans as a form of currency, made a bitter drink from it. In 1519, after tasting *cacahuatt*, a beverage enjoyed by Montezuma II, the last Aztec emperor, Hernando Cortés brought the beans back to Spain, where the chocolate drink was heated and mixed with sugar. For years the nobility kept the formula a secret. Word of this coveted bean finally leaked out in the mid-1600s, and soon European chocolate houses were as popular as coffee houses. The first chocolate candy was made in the early 19th century, when a Dutch chemist, Conrad Van Houten, invented a cocoa press that let confectioners mix cocoa butter with sugar.

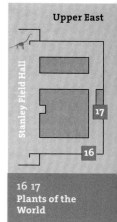

Upper East

16 17
Plants of the World

CACAO
Theobroma cacao

The California condor has had a rough 20th century. In 1940 there were fewer than 100 of the birds left. By 1987 the outlook was so grim that conservationists rounded up the remaining handful of condors for captive-breeding programs, leaving none in the wild. A common sight from British Columbia to Baja when the first European explorers arrived, this cousin of the Andean condor, with its wingspan of nearly 10 feet, has lost its habitat to humans: More people meant more sport hunters to kill condors, more egg collectors to steal their eggs, more pesticide to wind up in their food. Condors are a type of vulture, and they feed off the dead carcasses of large mammals, especially whales, deer, seals, elk, and bighorn sheep, which at one time were plentiful. But there is still hope for the California condor. Thanks to wildlife conservation organizations like the San Diego Wild Animal Park, which has an ongoing breeding program, there are more than 100 condors alive today, some of which have been released into the wild.

18 CALIFORNIA CONDOR
Gymnogyps californianus

LOCAL WOODLANDS FOUR SEASONS DIORAMA 19

Though it seems hard to imagine, when Carl Akeley completed his famous Four Seasons diorama, in 1902, the white-tailed deer was alarmingly scarce in Illinois and other states. Overhunted by humans, killed by wolves, and driven out of its natural habitat by farmers and real estate developers, deer populations dwindled. With these dioramas, Akeley wanted to preserve the deer for future generations who might never see them.

He also wanted to improve the way museums displayed animals. Until then, taxidermists typically stuffed the treated skins of dead specimens with straw and propped them up in unconvincing poses. For backdrops they would purchase heavy cloth leaves from millinery suppliers – in whatever styles were fashionable at the time – and tie them to limbs, usually without regard for habitat accuracy. A dedicated naturalist, Akeley, with the help of his wife, Delia, spent years working on this innovative display, much of the time late at night. For Akeley, the background details – the spring buds poking from the branches, a gray jay perched on the snow, the 17,000 wax leaves, each molded separately from real ones – were as important as the animals themselves, which he manipulated into lifelike poses using a structural support technique he invented. The Four Seasons, a breakthrough in museum displays, instantly became the standard to which all taxidermists aspired.

You can tell owls from other birds right away. They have large, broad heads with a ruff of feathers around the eyes called a facial disk. And owls are one of the only birds whose eyes face forward, giving them binocular vision, like humans. Unlike our eyes, however, their eyes do not move in their sockets, which is why the birds pivot their heads when watching moving objects. Most impressive is the owl's sense of hearing, which the facial disk heightens by channeling sound into the bird's large ear cavities. An owl hunting in total darkness can often locate and catch a mouse by listening to it rustle about on the forest floor. One of the largest of the world's 150 owl species, the great horned owl, found throughout Canada and the eastern United States, is 2 feet long and has a wingspan of almost 5 feet. This fierce predator nests in abandoned hawk or crow nests and hunts almost any small animal it can get its talons on — rabbits, skunks, snakes, and sometimes chickens.

GREAT HORNED OWL
Bubo virginianus
20

This bird is hanging upside down for a reason: He is trying to attract females. That is also why his wings are fluffed up like a peacock's and his two long tailfeathers are bent like arches. It is all part of his mating display. Like the 42 other types of birds of paradise, which live in New Guinea and nearby South Pacific islands, blue birds of paradise go to amazing extremes to lure a mate. Because the birds of paradise are typically fruit eaters, not insect hunters, and live in a habitat where fruit is plentiful, the males have time to "hang out" with their fellow birds. This male and his fellow courtiers assemble into what is called a lek to strut, sing, and flaunt their feathers. The most successful ones will mate with 40 or 50 females each year.

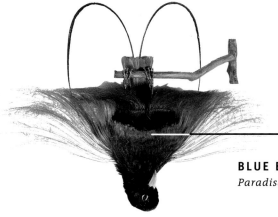

BLUE BIRD OF PARADISE
Paradisaea rudolphi
21

If you had a fairy armadillo in your palm, you could slide your finger underneath its hard shield all the way to the middle of the back, where a narrow ridge of flesh along the spine holds the pale pink shell in place.

Like its cousins, this armadillo is adept at burrowing for insects and has small, peglike teeth for chewing food.

Main West

20
North American Birds
21
World of Birds
22
World of Mammals

This is no ordinary armadillo. A native of the central deserts of Argentina, it is small, hairy, and flat at the rear, with leathery hind quarters.

22 PINK FAIRY ARMADILLO
Chlamyphorus truncatus

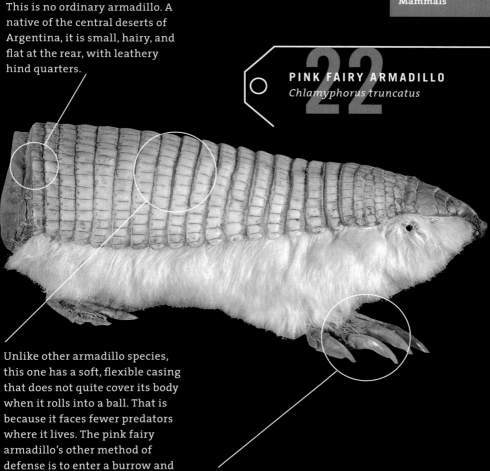

Unlike other armadillo species, this one has a soft, flexible casing that does not quite cover its body when it rolls into a ball. That is because it faces fewer predators where it lives. The pink fairy armadillo's other method of defense is to enter a burrow and close it off with its stubby rear, like a cork plugs a bottle.

Its hands are as big as its head.

23 Main West — Stanley Field Hall

23 World of Mammals

BLACK RIGHT WHALE SKELETON
Balaena glacialis

23

Hanging from the ceiling is the skeleton of a right whale, so named because early whalers considered it the right, or correct, whale to hunt. Known as a toothless, or baleen whale, it collects small fish and tiny shrimplike krill by filtering sea water through the rows of filaments – called baleen and made of a keratinous material much like human fingernails – suspended from its top jaw.

The two small bones floating unattached near the whale's tail fin are hipbones, the last traces of a bone structure that supported legs. Once land mammals, whales have evolved over millions of years to live in the water. That is one reason whales were able to grow so large, into the biggest creature ever to exist: An animal of such proportions would be helpless on land, but the water supports its enormous bulk.

Instead of hair to insulate them, like warm-blooded land animals, whales have layers of fat, called blubber, which also increases their buoyancy. Used by humans to make lamp and cooking oil, cosmetics, soap, and other products, this fat is one of the main reasons these whales are hunted.

Today the right whale, which can grow up to 60 feet in length, is an endangered species because of overwhaling. But international hunting regulations have helped many types of whales begin a comeback.

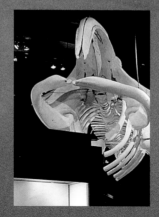

Black right whale skeleton suspended from Museum ceiling.

Main West

24 **Mammals of Asia**
25 **Messages from the Wilderness**
26 **Ground Level**

Ground

You can almost feel the cold wind blowing off the snow-covered peaks in this diorama. Or hear the clatter of the argali's hooves on the frozen, rocky slopes as the wild sheep descend. First described in the West by the Italian traveler Marco Polo (and sometimes called Marco Polo sheep), argali live in central Asia on the rolling plains between mountains, where they climb higher than any other mammals except mountain goats. Their most striking feature is their rack of horns, which can grow to more than 6 feet in length and can account for up to 10 percent of an adult male's weight. The rams use them to attract females and intimidate smaller males, often by challenging them to head-butting battles.

ARGALI
Ovis ammon
24

25 MEXICAN GRIZZLY BEAR
Ursus arctos

No one has seen a live Mexican grizzly bear since 1962. Abundant in the early 1900s when hordes of people began moving into northwestern Mexico's mountainous areas, these animals quickly lost their habitat, one of the main causes of animal extinction around the world. The bears were pushed from the shrubby foothills to the ponderosa pine forests of the Sierra Madre mountains. By the mid-1900s, loggers had cut down most of the pine trees, and many bears starved; ranchers hunted, trapped, and poisoned others. Today the ecosystem in this diorama no longer exists. Like a wilderness time capsule, this glass case is one of the only places in the world to see this unique creature.

26 BUSHMAN
Gorilla gorilla

When he was still alive, Bushman was the most famous gorilla in the country. Orphaned in Cameroon, West Africa, in the 1920s, he was brought to Chicago to live at the Lincoln Park Zoo. While there, he made regular appearances on Zoo Parade, a television program hosted by the zoo's director, Marlin Perkins. More than a million people a year visited Bushman.

When he turned 21, the mayor gave him a voter's registration card, joking that the 550-pound gorilla was probably as smart as most voters. Once, Bushman wandered out of his cage and into a nearby zoo kitchen. The zookeepers called the Chicago Police, but no one wanted to hurt Bushman. That's when one of Bushman's keepers remembered that he was afraid of snakes. The zookeeper ran to the reptile house and returned with a garter snake. As soon as Bushman spied the snake, he bolted back into his cage, closing the doors behind him.

In 1951, at the age of 23, Bushman died and was brought to The Field Museum, where he is still Chicago's favorite gorilla.

Main West

27 Stanley Field Hall

AFRICAN ELEPHANTS
Loxodonta africana

These two battling bull elephants are the work of the taxidermist Carl Akeley, who revolutionized the art of wildlife display during his 14 years at The Field Museum. As a young man working at Ward's Natural Science Establishment, in Rochester, New York, Akeley got his first big break in 1886, when he stuffed circus man P. T. Barnum's famous elephant, Jumbo. In mounting these two African pachyderms – collected by Akeley and his wife, Delia, on a 1905 trip to Kenya (she felled the larger of the two) – the inventive taxidermist built full-size frames using steel, wood, wire mesh, hemp, and plaster. Onto these he glued the hides, which he had cut into pieces, preserved with salt, and tanned. He then sewed together the edges and covered the seams in colored beeswax. If you look closely you can even see the bristly hair standing on the elephants' backs.

Akeley was also determined to mount his animals in dramatic, realistic poses. Here, one bull prepares to plunge its lone tusk into the other, which rears up on three legs. Akeley spent many months observing fauna in Africa, taking detailed notes and photographs on which to base his work. Despite a couple of brushes with death – he once killed an 80-pound leopard with his bare hands, and in 1912 he was crushed by a charging elephant – Akeley loved the African continent. First at The Field Museum and later at the American Museum of Natural History, in New York, he devoted more than three decades of his life to creating a lasting record of the animals of Africa, many of which were and are in danger of extinction.

Transporting elephant on rail car from Field Columbian Museum in Jackson Park to current Museum, 1920–21.

28 Daniel F. and Ada L. Rice Wildlife Research Station

THE LIONS OF TSAVO
Felis leo — 28

The year was 1898, and there was trouble at the Tsavo railroad camp. For weeks a pair of lions had terrorized workers, entering their tents at night, dragging men off, and devouring them. Three days after Colonel J. H. Patterson, a British engineer, arrived at the Kenyan outpost, the lions took another man. Patterson followed the trail. "The ground around us was covered with blood and morsels of flesh and bones," the colonel later wrote, "but the head had been left intact, the eyes staring wide open with a startled, horrified look." Frightened for their lives, the Indian and African workmen balanced their beds in trees and on top of water towers. Some slept in pits covered with railroad ties. Patterson was determined to stop the lions. One night, he and the camp's doctor planned an ambush, using cattle as bait and hiding with rifles in a box car. One of the lions lunged at the car's open door, and the men fired, missing the beast but frightening it away. Finally, Patterson built a makeshift scaffold, and from this unsteady perch shot and killed each of the two lions within a period of two weeks.

Colonel John H. Patterson with one of the Tsavo lions he shot, 1898.

Not all male lions have manes. Those that live in arid regions, like the Tsavo area of eastern Kenya, are often maneless. Lions in zoos typically grow the largest manes.

These lion skins are not in very good condition because for many years Col. Patterson used the skins as rugs. And before they were killed, the lions' hides were scarred by thorn-bush fences, called bomas, that workers erected around the camps. The Museum bought the rugs from Col. Patterson in 1924, for $5,000.

Before Patterson shot them, these man-eaters killed more than 100 people. The 1996 movie *Ghost and the Darkness*, starring Val Kilmer as Col. J. H. Patterson, was based on the reign of terror of these very lions.

If you are still feeling brave after staring into the eyes of these lions, look for the Man-Eater of Mfuwe on the ground level of the Museum. The largest man-eating lion on record, this fearsome creature killed at least six people in 1991 and was believed to be a demon by those he terrorized.

Crowned cranes
Balearica pavonina

Main West

29
Bird Habitats

30
Africa

MARSH BIRDS OF THE UPPER NILE RIVER 29

In 1954, members of an expedition to Uganda collected the brightly colored crane and 21 other bird specimens found in this case, which rounds out a series of African bird scenes from a rain forest, from a desert, and one of a colony of weaverbirds. The rarest bird in the diorama is the odd-looking whale-headed stork, the sole species in the family *Balaenicipitidae*, with its huge bill and feather crest. Somberly colored and solitary, its favorite food is lungfish. Also featured are a pair of pygmy geese, a crested grebe with a downy baby riding on its back, a swimming anhinga, and egrets, herons, a malachite kingfisher, and other birds. To help the exhibit planner recreate the scene from Lake Kyoga, on the Upper Nile, the expedition brought back color sketches and dried samples of plants for molding the models you see.

ELEPHANT DUNG BEETLE
Scarabaeidae family

30

Dung beetles, also known as scarabs, seek out steaming elephant waste, roll it into balls, and bury it, creating a warm, nutritious nursery for their eggs. Associating the rolling orbs of dung with the round sun traveling across the sky, ancient Egyptians deemed a type of dung beetle, the sacred scarab, a symbol of one of their most important deities, the sun god Re. Projections on the heads of the beetles, they believed, were emblems of the sun's rays. And because of what they perceived as a miracle of new life – young beetles emerging mysteriously from balls of dung – the scarab also came to symbolize resurrection and immortality. When people died, the Egyptians often removed their hearts and replaced them with large carved and often jeweled scarabs. Today the dung beetle also serves an important role as wildlife conservationist. By cleaning up after elephants, each of which can produce up to 200 pounds of dung per day, the beetles fertilize the savanna grasses. This, in turn, gives a boost to the food source of antelopes and other grazing animals.

Egyptian scarab

Powering a fine-grinding wheel with a foot treadle, a twentieth-century Chinese apprentice polishes a jade carving.

CULTURE

ANTHROPOLOGY at The Field Museum is all about what makes us human, our place in nature, our common concerns, and our different responses. Field Museum anthropologists build our collections and explore these issues through lab research and at field sites throughout the world.

The Field Museum's anthropology collections include more than 600,000 objects that document the diversity and accomplishments of humankind. Our collection of artifacts from the Pacific is the largest in the world. We also have world-class collections from each of the following regions: Africa, Europe, Asia, South America, Central and Middle America, and North America.

The **ANTHROPOLOGY DEPARTMENT'S** research is focused on the emergence and diversity of complex urban societies as well as the interrelationships between human groups and their environments. Curators today are conducting research in East Africa, the Southwestern United States, urban areas throughout the United States, China, Papua New Guinea, Mexico, Peru, and Amazonia.

"Walk a mile in my shoes," the saying goes. "Look at life from my point of view."

Depending on where people live, their history, and their creativity, the shoes they wear will be very different. To maneuver in deep snow, for instance, Eskimos wear broad, flat snowshoes made of tree branches and animal gut. Farmers in the Netherlands don heavy wooden shoes to keep their feet dry in the wet climate of the Dutch lowlands. And soccer players wear athletic shoes with metal or plastic cleats for traction. While the Living Together exhibit at The Field Museum uses many objects from around the world to help explain the reasons for cultural differences, the 135 shoes on display – including cowboy boots, silver wedding shoes from India, and Michael Jordan's hightops – are especially interesting symbols and are things we all can relate to.

31 Living Together

32 Inside Ancient Egypt

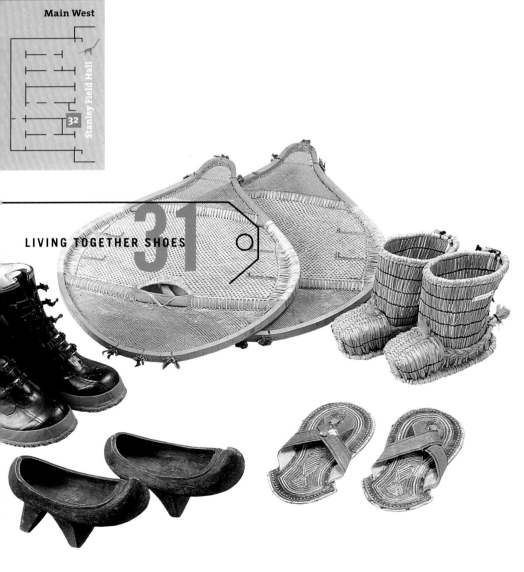

LIVING TOGETHER SHOES

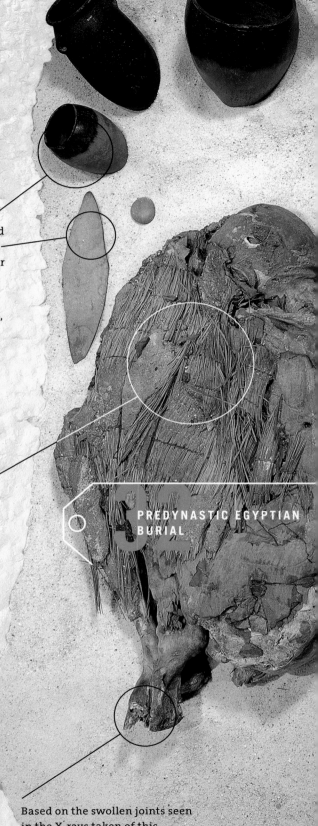

This Egyptian woman died more than 5,500 years ago, during the Naqada I period, 600 years before the first pharaohs, who developed elaborate methods of mummifying dead bodies. Her burial illustrates the beginnings of Egypt's rich funerary tradition. At the heart of this tradition was the belief that life did not end with death.

Surrounding her were items deemed essential for survival in the next life: black-and-red pottery jars of food and drink, and a stone makeup palette with pieces of malachite and galena for grinding into the familiar Egyptian eye shadow.

In order to survive in the afterlife, one's body had to remain intact. Before the Egyptians began embalming and wrapping corpses, they buried them in shallow, desert graves, which protected the bodies from the sun's decaying rays and allowed them to dry. Wrapped in a reed mat, this 50-year-old woman's body was placed in the fetal position, perhaps awaiting rebirth.

PREDYNASTIC EGYPTIAN BURIAL

Based on the swollen joints seen in the X-rays taken of this naturally mummified body, scientists can tell that the woman suffered from arthritis.

Main West

Stanley Field Hall

33
Inside Ancient Egypt

EGYPTIAN MASTABA 33

As ancient Egyptian culture became more complex, so too did Egyptian funerary practices. By the time of the early dynasties, instead of dry sand graves, wealthier families had begun to build long, low mausoleums called mastabas, a word that means "bench" in Arabic. In these multiroomed "mansions of eternity," as the Egyptians called them, the *ba*, or "wandering spirit," of the deceased could live free and unconstrained. Reconstructed using two of its original six rooms, the 4,400-year-old mastaba of Unis-ankh, son of a 5th Dynasty pharaoh named Unis, now stands in The Field Museum.

Continued from the predynastic days was the belief that the deceased's body must remain intact. But without the desiccating sand of the desert graves, Egyptians had to develop other methods of preserving corpses. This led to the technique of mummification. A common practice was to remove the rapidly decaying organs – the stomach, intestines, liver, and lungs – and to dry the body using natron, a natural salt preservative found in the bottoms of shallow lakes. Wrapped and placed in a coffin, the mummy was then laid to rest in a mastaba or some other rock-cut tomb. Only the pharaohs and their queens were buried in pyramids. The mastabas were used mainly by nobility.

Inside the mastaba is a false door, the meeting place of the living and the dead. Unis-ankh's *ba* would have passed through this stone to visit his *ka*, or earthbound spirit, to receive offerings of food and drink.

No one knows for sure what is inside this beautiful, perfectly preserved inner coffin. In the century since it was donated to the Museum, it has never been opened – to do so would no doubt damage it – nor has it been X-rayed. According to hieroglyphic inscriptions on its surface, it contains the remains of Chenet-a-a, the "lady of the house," who lived between 945 and 712 B.C., during the 22nd Dynasty. It is made of cartonnage, a papier-mâché-like material made from glued papyrus sheets that are plastered, painted, and varnished. Until Chenet-a-a's mummified body was slipped inside, the inner coffin was left open at the back. Then, it was sewn up and a sheet of cloth sealed the slit. The name of Chenet-a-a's father is written on the wooden outer coffin, at the time a common practice among Egyptian nobility.

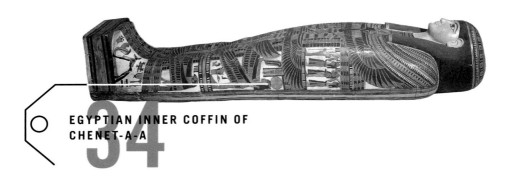

EGYPTIAN INNER COFFIN OF CHENET-A-A
34

Egyptologists hold two theories about the origin of this vessel, one of six boats buried near the tomb of the Egyptian pharaoh Sen-Wosret III. One is that when the 12th-Dynasty pharaoh, who lived on the east bank of the Nile River, died, in 1842 B.C., his followers used the boats to carry funerary offerings – food, beer, jewelry – across the river to his tomb. The other theory is that the boats were meant to transport the dead king in the afterlife across heaven with Re, the sun god. In 1950, when scientists at the University of Chicago were developing the carbon-14 dating system, they borrowed a plank out of this boat. Because they knew the date the king died from the stories carved in his tomb, they were able to prove that the dating system worked.

EGYPTIAN FUNERAL BOAT OF SEN-WOSRET III
35

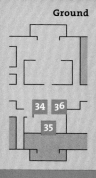

Ground

34 35 36
Inside Ancient Egypt

36 EGYPTIAN CAT BRONZE

In ancient Egypt, when a pet cat died, its owner shaved his or her own eyebrows as a sign of mourning. Cats were so sacred, in fact, that if you killed one, you risked being put to death. That's because the Egyptians considered cats to be incarnations of the goddess Bastet, daughter of Re, the sun god. Bastet was a benevolent, loving deity. She protected people against evil in the same way that a cat – called a *miu* in the Egyptian language – cleared snakes and rodents from a farmer's fields. The earliest known feline art dates to the 1st Dynasty, or about 3100 B.C.

This exquisite bronze cat is most likely from the 26th Dynasty, between 656 and 525 B.C. Molded into the statue are a sacred scarab on her head, an *uskh* necklace, and a protective eye amulet on her breast, all inlaid with gold, silver, and copper. Inside the hollow cast is a bundle – seen only in X-rays taken of the statue – that Egyptologists believe holds the remains of a dead cat. That's not surprising, considering the fact that Egyptians regularly embalmed, mummified, and buried the animals in coffins, just as they did with nobles.

Main West

37 38 39
Africa

37 BAKONGO NKONDI FIGURE

When two people in the United States cannot resolve a dispute, they sometimes go to a mediator, a third party who hears evidence from each side and settles the case. For the Bakongo people – who live in the countries of the Congo, the Democratic Republic of the Congo, and Angola – one custom for resolving a dispute is to go to a *nganga*, a "wise man" who decides the matter with the help of ancestral spirits. He does this using a *nkondi*, like the 19th-century figure on display at The Field Museum. Performing a series of dramatic rituals, the *nganga* activates the *nkondi*, which means "agreement" or "consensus." The figure becomes a symbol of empowerment that validates the *nganga*'s decision.

Each time the *nkondi* figure is used to settle an argument, the *nganga* hammers a piece of metal into the statue. Over time, as more and more nails are pounded into the figure, a *nkondi* like this one becomes a powerful symbol of the wisdom that determines right and wrong.

What is now the state of Edo in Nigeria was until 1897 a powerful African kingdom ruled by men known as *obas*. (*Obas* still exist, but today they no longer have political authority under the Nigerian government.) For many generations, royal law dictated that the *oba*'s mother – the Queen Mother – be put to death to prevent her from undermining her son's power. But about 500 years ago, this practice changed when an Edo prince named Esigie became king. Instead of killing his mother, he sent her away, vowing never to see her again. Since then, whenever a man is made *oba*, the Queen Mother's life is spared, but she is exiled to her own palace. When the Queen Mother dies, the *oba* dedicates an altar to her memory, the centerpiece of which is a bronze memorial head, like this one, which dates to the 19th century.

38 BENIN QUEEN MOTHER MEMORIAL HEAD

The Tuareg, who live mainly in Algeria, Mali, and Niger, are the largest group of nomadic people in the Sahara Desert. They typically herd camels, goats, sheep, and cattle, striking their tents and moving periodically in search of pastures for their livestock. Outfitted with a saddle over its hump, the camel is the Tuareg's main mode of transportation, as important to them as the pickup truck is to many farmers in the United States. To carry their belongings, the Tuareg make brightly colored leather-and-yarn saddlebags, like this one, from the mid-20th century.

39 TUAREG SADDLEBAG

40 North American Indians

Today's natural history museums are dynamic places where different peoples share ideas and stories. Take, for instance, the ongoing relationship between The Field Museum and the Pawnee, a group of North American Indians based in Oklahoma. When the idea arose more than two decades ago to build an Indian earth lodge as a center of learning and interactivity, Museum staff first looked to the Pawnee for permission and consultation. The result of the collaboration is one of the few full-size earth lodge reproductions in the world.

PAWNEE EARTH LODGE

Pawnee Earth Lodge at the Louisiana Purchase Exposition, St. Louis, Missouri, 1904.

The Pawnee of the mid-19th century, who lived on the Great Plains in present-day Nebraska, built their lodges out of available materials: cottonwood poles for support, willow-branch lathe work for the frame, and a covering of mud and sod. A large round building with a domed roof – a shape that symbolized their view of the universe – the Pawnee house always faced east, out of which Mars, the morning star, rose. On the western side of the lodge – in honor of Venus, the evening star – stood the sacred area. This corner, accessible only to the priest, usually contained a buffalo skull, a corn pit, and a sacred bundle, with its generations-old religious symbols. Two extended families totaling up to 50 people could live in one lodge.

The Pawnee considered the stars and planets sacred, and cosmology influenced where different family members slept. The evening star symbolized femininity and fertility, so the teenage girls bunked near the sacred area. The grandmothers slept opposite them, near the doorway. When the elderly women died, their spirits could leave and travel the galaxy, becoming stars and joining their ancestors. Twice a year – as soon as the crops came up in the summer and again in the winter – the families would leave their lodges and hunt buffalo on the plains, camping in animal-hide teepees until they returned.

**41 42
North American Indians**

HOPI GREAT HORNED OWL KACHINA DOLLS 41

Though they are colorfully painted and occasionally feature feathers and other adornments, these dolls are not playthings. They are used to teach Hopi children about the spiritual beings called kachinas. The Hopi believe that kachinas once lived among them, dancing and bringing pleasure and goodness to the lives of the Indians. But the kachinas took offense, legend has it, at human evil and corruption and moved to Arizona's San Francisco Peaks, where they remain half the year. During the other half, beginning in late February, they return to the village for ceremonies. There are many different kachinas, representing hundreds of spirits – such as the hummingbird kachina, the bean-sprout guard kachina, and the kind and gentle kachina – and their dances are prayers for rain, healthy crops, and other blessings. On the days of the ceremonies, kachina dancers present the children with the dolls, typically carved from a cottonwood root by one of the children's relatives.

These two dolls represent the great horned owl kachina. The one above dates to 1992, and is one of the first types of dolls given to toddlers; the other, made in the 1900s, would have been used as a learning tool by older children.

The Apache Indians, who settled in the southwestern United States, are known for their exceptional basket weaving. By bending slender willow, cottonwood, or sumac twigs into a three-rod foundation, and then wrapping the foundation piece with twig or bark sewing splints, Apache women create rigid coil baskets in a variety of shapes and styles. The weavers don't dye their coil baskets. Instead, they use a desert plant called Devil's Claw for black strands, and they peel and prepare the inner bark of the yucca root for red ones. The 19th-century Apache stored grain and other foods in most baskets, which were better suited to their nomadic way of life than fragile pottery. They coated others with a mixture of ground leaves and piñon pine pitch for use as water bottles. Although tradition and utility dictate the general appearance of the baskets, each weaver expresses herself as an individual artist, which is why there are so many different beautiful basket styles.

42 APACHE JAR-SHAPED COIL BASKET

Jicarilla Apache (New Mexico) woman weaving at the Louisiana Purchase Exposition, St. Louis Missouri, 1904.

There is no written record and little in the way of oral history passed down about the Hopewell Indian culture, which originated in what is now the state of Ohio and flourished from 100 B.C. to 400 A.D. Almost all of what anthropologists know about this people has been pieced together from archeological findings, like this exquisite bird claw, from an extensive series of earthen mounds, many of which are in Ohio. ("Hopewell" was the name of the man who owned the farm where some of the excavations took place.) The largest mound is 500 feet long, 180 feet wide, and 30 feet high. Some mounds were used for burying the dead and others as platforms for temples and houses. Known for their beautiful work, Hopewell artists cut, sculpted, and engraved stone, metals, and bone, often in the shapes of birds, fish, bears, people, snakes, and other things from nature. They developed an extensive trading network, obtaining alligator and shark teeth from the Gulf coast, obsidian from the Rocky Mountains, and silver from the Great Lakes region. The mica used in this cutout probably came from the southern Appalachian Mountains.

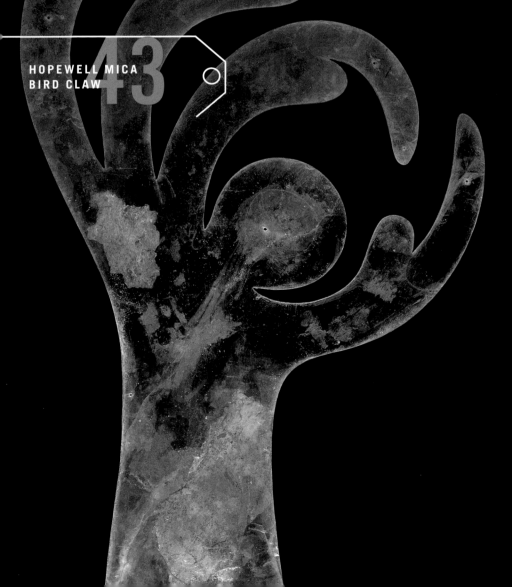

HOPEWELL MICA BIRD CLAW 43

This late 19th-century carving of a man holding two land otters was once used to catch fish. The Tlingit, Haida, Kwakuitl, and other peoples of North America's Northwest Coast carved elaborate designs into their wooden hooks in an attempt to impress the spirit of the halibut enough so that it might allow itself to be caught. Like the dances and prayers the Indians make to the animal spirits, the carving ritual demonstrates the respect they show to their prey. And the halibut, one of the world's largest fishes (they can grow to more than 400 pounds), is an important food source for the people living away from the salmon-rich rivers.

The southern Indian peoples would steam-bend fir branches into U-shaped hooks. Those of the north would create a V-shaped hook by binding two straight pieces – one carved, the other containing the barb – using cedar bark or animal sinews. Baited with squid or octopus, the hooks sat barb-up on the bottom awaiting an "impressed" halibut.

43
Ancient North America and Mesoamerica

44
Eskimos and Northwest Coast Indians

44 TLINGIT HALIBUT HOOK

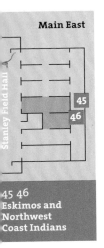

**45 46
Eskimos and
Northwest
Coast Indians**

Among the Indians of the Northwest Coast of North America, the "copper," a handcrafted metal plaque, was once the most valuable and treasured piece of personal property – and to destroy one was the most powerful gesture a chief could make to earn respect. The breaking of coppers was one part of certain "potlatch" ceremonies, three- or four-day competitive celebrations that pitted the leader of one clan against the leader of another. The goal was to give away food, blankets, and other goods in order to demonstrate one's excessively superior, and thus disposable, wealth.

Hammered into the shape of a flared shield and decorated with paint, a copper, like this 19th-century piece, was worth thousands of blankets, the Indian's form of currency. At a potlatch, in front of a great audience, the chief would ceremoniously break the copper into five pieces, beginning with the upper right-hand corner, and give the pieces to his rival. An even more impressive sign of excess wealth was to throw a copper into the sea. The coppers often had names, such as Making the House Empty of Blankets, a reference to the copper's power, and Making Ashamed.

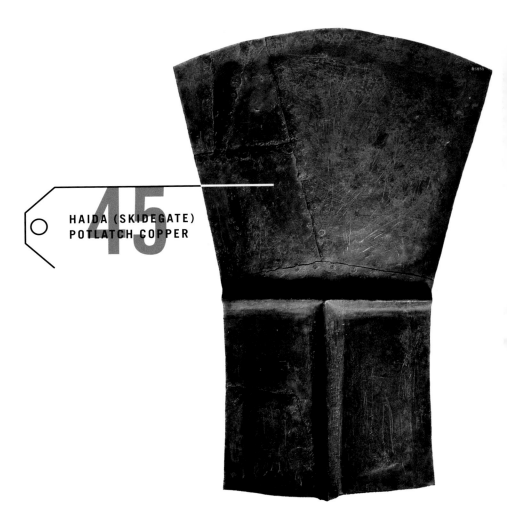

45 HAIDA (SKIDEGATE) POTLATCH COPPER

POINT HOPE SHAMAN MASK 46

Largely due to where they live – in the cold, harsh tundra, often above the Arctic Circle – Eskimos developed a more spare material culture than their southern Indian neighbors. Their ceremonial masks, rarely painted or embellished, reflect this austerity. The Eskimos believe that spirits inhabit all things and that supernatural forces manipulate their lives. They wear masks during ceremonial dances in order to placate the spirits – who have power over the Eskimos' food supply, for example – and to bind the human world to the realm of myths. That is why so many Eskimo masks resemble animals, like whales, seals, and other creatures of the hunt. Others look like human faces because the spirits also manifest themselves in human form. While ordinary people make and use masks, it is the shaman who has the most influence with the otherworldly. He translates his intense visions into powerful masks, like this one, which dates to the late 19th century and is thought to represent the moon.

	Main East
	47
49	
	48

**47 48
Eskimos and Northwest Coast Indians**

**49
Ancient North America and Mesoamerica**

This 19th-century carved-cedar mask can be opened and closed mechanically by the person wearing it, revealing different faces. Made by the Kwakiutl Indians, of the Pacific coast of Canada, it is typically worn by a male dancer in one of the group's many ceremonies. Transformation masks sometimes feature an outer human face and an inner animal face. Like the Eskimo peoples, the Indians of the Northwest Coast maintain a close, respectful relationship with animals, especially the ones they rely on for food. But unlike the simple Eskimo masks – often flat and unpainted – the masks of the Northwest Coast Indians are quite elaborate. The bigger, bolder, and more brightly colored a mask is, the more respect the owner enjoys from others in his society.

**KWAKIUTL
TRANSFORMATION MASK**

47

This towering, late 19th-century bear statue once served as the entrance to a house. Acquired for Chicago's 1893 World's Columbian Exposition – the international event that led to the creation of The Field Museum – it has been a highlight of the Pacific Coast collection ever since. House posts typically stand in front of or near the threshold. Like totem poles, they contain carved symbols of the family's lineage, similar to the European coat of arms. Exhibiting such a statue is a demonstration of pride and a way to show respect to the spirit of the grizzly bear. In certain cases, there are taboos against a clan hunting its symbolic animal. In this way, the totemic system acts as a means of environmental conservation.

48 BELLA COOLA GRIZZLY BEAR HOUSE POST

Many indigenous cultures in Mexico and Central America worshipped Quetzalcoatl, considered the god of culture and learning, among other things, and represented by a plumed serpent. The god's name is a combination of the word *quetzal*, a blue-and-green bird (the national bird of Guatemala), and *coatl*, which means "snake" in the Nahua language. This diorite sculpture, probably carved between 1250 and 1519, features two serpents – one with a human face in his jaws – and two people piercing their ears with bone skewers. Some believe that when Spanish conquistador Hernando Cortés (1485–1547) advanced on the Aztecs, their ruler, Montezuma II, did not fight because he feared Cortés might be Quetzalcoatl. Cortés ultimately captured Montezuma and defeated the Aztecs.

49 AZTEC DIORITE BOULDER WITH QUETZALCOATL FIGURE

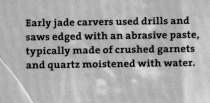

Early jade carvers used drills and saws edged with an abrasive paste, typically made of crushed garnets and quartz moistened with water.

50 CHINESE JADE JAR

While many cultures have carved jade, Chinese craftspeople elevated the art form to its greatest height. In China, jade traditionally has been as precious as gold, prized not just for its market value but for its beauty, rarity, antiquity, and ritual meaning. Writing 2,500 years ago, the philosopher Confucius compared five qualities of jade with human virtues: Like charity, it has a bright but warm luster; like rectitude, its translucency reveals a strong inner character; like wisdom, jade makes a pure and penetrating sound when struck; like courage, it can be broken but not bent; and like fairness, its sharp angles injure no one. Made in the 18th century for the Qianlong emperor, this exquisite, 281-pound jar was placed in the Imperial Palace in Beijing, where it became known as one of the "Eight Extraordinarily Large Pieces." Because extended exposure to bright light can alter the color of jade, the Museum's jade room is kept dimly lit.

Guanyin is one of the most important Buddhist deities. Considered a bodhisattva, Guanyin usually appears as a male in Tibetan and early Chinese art. In most late Chinese art, Guanyin is shown as female and becomes the goddess of mercy. Followers turn to Guanyin for fertility help and protection from danger. This rare figure is one of the finest examples of Yuan dynasty (1279–1368) ceramic sculpture, made in the famous porcelain-producing city of Jingdezhen, in eastern China. Although only about a foot high, the piece has a monumental feeling, exhibiting a wonderful combination of delicacy and power.

51 CHINESE PORCELAIN OF GUANYIN

Upper East

50
Hall of Jades

51
Grainger Gallery

52
China

Upper West

This painting was modeled on an early medieval European work called *Salus Populi Romani* ("The Health of the Roman People"), from the Church of Santa Maria Maggiore, in Rome. The unknown artist most likely saw a reproduction that had been brought to China, perhaps by Catholic missionaries, who first went there in the late 16th century. The most famous of these missionaries was Matteo Ricci, a Jesuit priest of legendary intellect. The emperor himself was impressed by Ricci's astounding memory, and certain members of the Chinese elite admired him enough to adopt his faith. Although Buddhism, Taoism, and Confucianism are the chief religions in China, a small Christian minority has worshipped there for the past 400 years.

52 CHINESE CHRISTIAN MADONNA AND CHILD

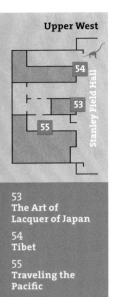

Upper West

53
The Art of
Lacquer of Japan

54
Tibet

55
Traveling the
Pacific

What does this beautiful Japanese container have in common with poison ivy? Known as inro, it is made of lacquer, which comes from the toxic sap of the urushi tree (*Rhus vernicifera*), a close relative of poison ivy (*Rhus toxicodendron*). Craftspeople collect, filter, and purify the sap, and then apply the lacquer in thin layers. Once it has hardened, it no longer causes skin rashes. But because each layer can take weeks or months to dry, an inro of 20 layers can take several years to complete. While the Japanese were not the first to work with lacquer – the Chinese used it to waterproof wood 3,500 years ago – they perfected the painstaking decorative technique, and by the 1500s Japanese artists were producing some of the world's finest lacquer boxes, bowls, tables, and tea trays. They were especially talented at making inro, which originally were used as medicine cases and later were hung from waistbands as personal adornment. Most of the pieces in The Field Museum's exhibit date from the Tokugawa period (about 1600 to 1850).

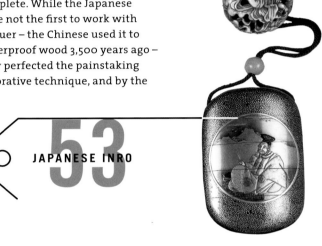

53 JAPANESE INRO

This is an 18th-century statue of Tibet's patron saint, Chenrezi. Considered the god of compassion, the figure represented here in silver – with ornamental gold, pearls, turquoise, and coral – according to Tibetan legend, was once an ordinary person. On his way to become a Buddha and enter Nirvana, Chenrezi witnessed the weeping and wailing of the people on earth who mourned his death. The sight caused him such pain that he vowed to devote himself to saving suffering people. Instead of becoming a Buddha, he became what is known as a bodhisattva, or a kind of Buddha-to-be. Today Chenrezi is embodied in the Dalai Lama, the spiritual leader of Tibetan Buddhism.

54 TIBETAN SILVER DEITY, CHENREZI

The outrigger canoe developed by the people of the South Pacific is a prime example of how environment can influence technology. In a region made up of thousands of small islands, the outrigger is a way to overcome the barrier created by the vast expanses of water. The floating arm steadies the canoe, and the sail propels it swiftly across the water, even in rough seas. The outrigger canoe has helped the people of these coral atolls and other islands fish, trade, wage war, and simply communicate with their neighbors. Before building materials such as plywood, caulk, rope, nails, and nylon became available through trade and commerce with Europeans, islanders used local materials: driftwood logs for hulls, coconut fiber for lashing parts together, and breadfruit sap mixed with the aerial roots of the pandanus tree to make a natural sealant.

55 JALUIT ATOLL DIORAMA
Marshall Islands

To the Maori people of Aotearoa (New Zealand), this ornately carved house is a sacred symbol of community pride and a place of reflection and social gathering. Even as it stands today – inside the Museum – Maori customs govern the use of this meeting place and of the *marae*, or area surrounding it. Those being formally welcomed should arrive in silence, and the *marae atae*, or area directly in front of the house, must be kept clear at all times.

Known as Ruatepupuke II, the house was built in 1881 in the town of Tokomaru Bay, on the north island of New Zealand. The Field Museum acquired it from a German collector in 1905. In 1992 Maori elders and artists and Field Museum staff restored the house. The people of Tokomaru Bay gathered kakaho reeds and kiekie strands – the same materials their forefathers used – to weave the new tukutuku panels that line the interior walls. At dawn on March 9, 1993, a delegation of 15 Tokomaru Maori formally reopened Ruatepupuke II. Maori people traveling through the Midwest continue to use it as their spiritual outpost away from home.

56 RUATEPUPUKE II
A Maori Meeting House

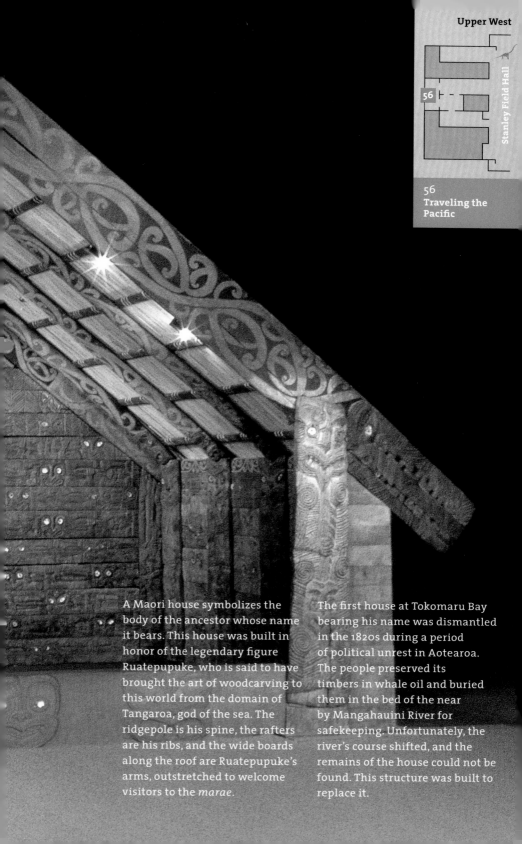

Upper West

56 Traveling the Pacific

A Maori house symbolizes the body of the ancestor whose name it bears. This house was built in honor of the legendary figure Ruatepupuke, who is said to have brought the art of woodcarving to this world from the domain of Tangaroa, god of the sea. The ridgepole is his spine, the rafters are his ribs, and the wide boards along the roof are Ruatepupuke's arms, outstretched to welcome visitors to the *marae*.

The first house at Tokomaru Bay bearing his name was dismantled in the 1820s during a period of political unrest in Aotearoa. The people preserved its timbers in whale oil and buried them in the bed of the near by Mangahauini River for safekeeping. Unfortunately, the river's course shifted, and the remains of the house could not be found. This structure was built to replace it.

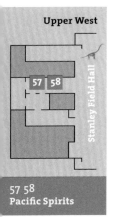

Upper West

57 58
Pacific Spirits

This rare, early 19th-century statue survived a time of religious and social upheaval, when most holy icons were destroyed. When Kamehameha I, founder of the Kingdom of Hawaii, died in 1819, power passed to his son, Liholiho, who became known as Kamehameha II. One of the young ruler's first acts was to abolish the existing religion, a rich system of beliefs that included multiple gods and goddesses, many orders of priests, and strict taboos. There were gods of creation, harvest, and war, and specific gods worshipped by particular tradesmen, like fishermen and carvers.

All that changed with the arrival of Christian missionaries and with Liholiho's decree. Temples were destroyed and wooden statues of deities were burned. This god figure, however, was one of a pair salvaged by the naturalist Andrew Bloxam when his British ship, the H.M.S. *Blonde*, visited Hawaii in 1825. Bloxam found the figures in Hale-o-Keawe, a royal mausoleum, beside the bundled bones of former high chiefs. The statues are thought to have been personal gods watching over the deified remains of these prekingdom rulers, who happen to have been Kamehameha's ancestors. That may be the only reason this particular mausoleum was spared. The other god figure is in Hawaii's Bishop Museum.

57
POLYNESIAN GOD FIGURE
Bloxam Statue

The people of Vanuatu, an archipelago of islands in Melanesia, believe that earthly life and the afterlife are continuous. In fact, villagers often show respect to a living man who has achieved a high social rank by calling him *temes*, which means "dead man." When a revered villager dies, he takes his honored place among the ancestors. Using the dead man's skull, his friends and relatives create a stick figure, called a *rambaramp*, in his likeness, which they display in the men's meeting house. They paint the *rambaramp* in a way that reveals the deceased's social rank, and they treat it like a living person. This one, which dates to the late 19th century, is from the Vanuatuan island of Malekula.

MELANESIAN RAMBARAMP
58

59 Pacific Spirits

60 Searle Lounge, West, Living Together Satellite

To the people of New Ireland, an island near New Guinea, in Melanesia, the pig has long been an important animal. At every stage of the sacred funeral ceremonies known as *malanggan*, villagers exchange pigs as payments to the deceased relatives to ward off harm from the spiritual forces, and they eat them during *malanggan* feasts. At the ceremony's finale the patron climbs atop a pile of sacrificed pigs to eulogize his dead relative. He then carves up and passes out pork for guests to take home with them. Dancers at the New Ireland *malanggan* and other rituals wear masks like this one (inset), which dates to about 1910.

MELANESIAN PIG MASK

Man wearing pig mask, New Ireland, circa. 1900.

Malvina Hoffman (American, 1885–1966), who sculpted these busts, learned – partly under the tutelage of Auguste Rodin, in Paris – how to shape clay, bend iron, and saw wood. Rodin also taught her about aesthetics: he often took his pupil to the Louvre Museum just before closing, where by the light of a candle he would show her the strong, smooth lines of the great statues. In 1930 Field Museum president Stanley Field commissioned a series of 104 life-size portraits of people from the world's cultures. The bronze busts you see here – of a Zulu woman, from southern Africa, and a Padaung woman, from Burma – are two of the several dozen sculptures still on display. To complete her assignment, Hoffman spent five years traveling the globe, meeting people from other cultures, and capturing in bronze the telling details about their customs and personalities. These two portraits help to illustrate how different cultures hold different standards of beauty.

60 MALVINA HOFFMAN ZULU WOMAN AND PADAUNG WOMAN

Padaung

Zulu

The Field Museum
Roosevelt Road
at Lake Shore Drive
Chicago, Illinois 60605-2496
[312] 922 9410
http://www.fieldmuseum.org

© 1998, 2000 by The Field
Museum
All rights reserved.
ISBN 0-914868-20-9

Printed on recycled paper ♻

Project Manager: Sophia Shaw
Editor: David Schabes
Photo archivist: Nina Cummings
Design and typesetting:
studio blue, Chicago
Color separations:
Professional Graphics, Inc.
Printing and binding: Arnaldo
Mondadori Editore, S.p.A.

The following people generously provided assistance in the preparation of this book:
Mark Alvey, Ruth Ingeborg Andris, Maria Arias, Gretchen Baker, Mary Ann Bloom, Rosaura Boone, Bennet Bronson, Paul Brunsvold, William Burger, Marla Camp, Joan Connor, Corrie Coughlin, Peter Crane, Annie Crawley, Laura and Tony Davis, Richard Faron, John Flynn, Sharleene Frank, Laura Gates, Lance Grande, Charles Grantham, Jonathan Haas, Melissa Hilton, Dennis Kinzig, Janice Klein, Chapurukha Kusimba, Peter Laraba, Leonore Leavit, Jill Mandler, Michelle Miller, Debra Moskovits, Gregory Mueller, Harry Nelson, Gertrude Novak, Clarita Nunez, Marie Orendach, Philip Parrillo, Bruce Patterson, Maureen Ransom, Jay Savage, Florence Selko, William Simpson, Bob Spieler, William Stanley, Douglas Stotz, Daniel Summers, John Terrell, James VanStone, Robert Vosper, Meenakshi Wadhwa, John Wagner, Robin Wagner, Alaka Wali, John Weinstein, David Willard, Frank Yurco.

Credits: All photographs © by The Field Museum, unless otherwise noted. Photographers' names are abbreviated: JB=James Balodimas; CC=Charles Carpenter; LD=Linda Dorman; GP=George Papadakis; RT=Ron Testa; MT=Michael Tropea; SAW= Sophia Anastasiou Wasik; JW=John Weinstein; DAW=Diane Alexander White, MW=Mark Widhalm.

Title page: © 1998 Photodisc, Inc.; **intro**: GN88071, 44672 (inset); **nature**: GEO4024; **1**: © 1998 Photodisc, Inc., CSGEO62911 (inset); **2**: GEO84618 RT; **3**: MT1; **4**: GEO79617, GEO79605 (inset); **5**: GEO84975P RT; **6**: GEO85820 JW; **7**: CK5T RT; **8**: GEO85858.12 JW; **9**: GEO84989 RT; **10**: 88688.35ac JW; **11**: GEO86160.3c JW; **12**: 89201.5c MW, Z94275C JW (inset); **13**: MT2; **14**: MT3; **15**: B79563, B83024 RT (inset); **16**: B83007 RT; **17**: MT4; **18**: Z93874 JW; **19**: Z93889 JW, Z93886.1 JW (inset); **20**: MT5; **21**: MT6; **22**: MT7; **23**: © Flip Nicklin, Minden Pictures, GN86102.2 JW (inset); **24**: MT8; **25**: Z22 RT; **26**: Z2 RT; **27**: GN86833 JB, CSGN74660 (inset); **28**: Z93658, GN87710 GP (inset); **29**: Z93624 RT; **30**: Z88196 JW, A110664 (scarab); **culture**: A86439; **31**: A113131.3 JW; **32**: MT9; **33**: A111519 RT, A111511 RT; **34**: MT10; **35**: 8179; **36**: A111088 DAW; **37**: A109327 DAW; **38**: A113376 JW; **39**: A112459 JW; **40**: CSA15648 CC, A106901 RT (inset); **41**: A113374.2c JW (left), A113373 JW (right); **42**: CSA15152 CC, A113375 JW (inset); **43**: A110016 RT; **44**: A113378 JW; **45**: A113379 JW; **46**: A108555 RT; **47**: A108352 RT; **48**: CSA73813; **49**: A857 RT; **50**: A86440, MT11 (inset); **51**: A113377 JW; **52**: MT12; **53**: A110381 SAW; **54**: A100790; **55**: A111635 JW & JB; **56**: A112518 DAW & LD; **57**: MT13; **58**: A111272 JW; **59**: CSA27340, A111576 JW (inset); **60**: MH12 (left), MH90 (right).

Anthropology accession numbers: **32**: 31736; **33**: 24448; **34**: 30005; **35**: 30286; **36**: 31760; **37**: 91300; **38**: 8262; **39**: 222077; **41**: 44011; **42**: 68612; **43**: 110131; **44**: 179860; **45**: 85038; **46**: 53459; **47**: 19166; **48**: 18634; **49**: 48102; **50**: 235936; **51**: 119332; **52**: 116027; **53**: 284830–2; **54**: 32906; **55**: 236809; **56**: 143961; **57**: 272689; **58**: 37720; **59**: 138855; **60**: MH12 (left), MH90 (right).

Inside front cover map courtesy Naughton + Associates, Inc.